READING POWER

Earth Rocks!

Metamorphic Rocks

Holly Cefrey

The Rosen Publishing Group's
PowerKids Press™
New York

Published in 2003 by The Rosen Publishing Group, Inc.
29 East 21st Street, New York, NY 10010

First Edition

Book Design: Mindy Liu

Photo Credits: Cover, p. 21 © David Muench/Corbis; p. 5 © Frank Lane Picture Agency/Corbis; pp. 6 (inset), 10 Mindy Liu; pp. 6–7 © Digital Vision; p. 9 © Kevin Schafer/Corbis; p. 11 © Roger Ressmeyer/Corbis; p. 13 (top) © George D. Lepp/Corbis; p. 13 (middle right) © Lester V. Bergman/Corbis; p. 13 (middle left) © E.R. Degginger/Animals Animals; p. 13 (bottom) © Galen Rowell/Corbis; pp. 14, 15 © Alfred Pasieka/Science Photo Library; p. 17 (top) © Lester V. Bergman/Corbis; p. 17 (bottom) © Kaj R. Svensson/Science Photo Library; p. 18 © Tiziana and Gianni Baldizzone/Corbis; p. 19 © Bob Krist/Corbis

Library of Congress Cataloging-in-Publication Data

Cefrey, Holly.
Metamorphic rocks / Holly Cefrey.
  p. cm. — (Earth rocks!)
Summary: Describes some of the properties and uses of metamorphic rocks.
Includes bibliographical references and index.
ISBN 0-8239-6466-3 (library binding)
1. Rocks, Metamorphic—Juvenile literature. [1. Rocks, Metamorphic.]
I. Title.
QE475.A2 C44 2003
552'.4—dc21

                                        2002000121

Contents

Metamorphic Rocks!

Metamorphic means "change of form" in the Greek language. Metamorphic rocks were all once other kinds of rocks. Over millions of years, these other kinds of rocks were changed to become metamorphic rocks. Metamorphic rocks can be made in different ways.

The Fact Box

There are two other kinds of rocks on Earth. They are igneous rocks and sedimentary rocks. Igneous rocks are made from lava and magma. Sedimentary rocks are made from sediment.

Metamorphic rocks are found all over the world. This metamorphic rock is in Germany.

One way metamorphic rock can be made is by heat. Earth is made of many layers of rock. Some of that rock is magma. Magma is melted rock. The heat from the magma can change other kinds of rock into metamorphic rock.

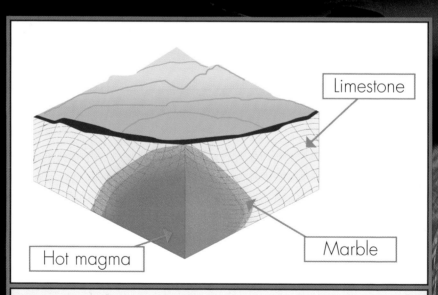

Limestone

Marble

Hot magma

Rocks melt at temperatures between 392 and 2,012 degrees Fahrenheit (200 and 1,100 degrees Celsius). Heat from the magma inside Earth can change limestone, a sedimentary rock, to marble, a metamorphic rock.

The Fact Box

Very hot water inside Earth can also change some rocks into metamorphic rocks.

Metamorphic rocks can also be made by pressure deep inside Earth. This pressure works with heat inside Earth to make metamorphic rock.

Gneiss (NICE) is made deep below Earth's surface. It can only be seen when the top rock has been worn away or the bottom rocks have been pushed up to the top.

Metamorphic rocks are also made when parts of Earth are pushed together. Earth's surface is made of huge, solid pieces called plates. Plates move very slowly. When two plates hit one another, rocks inside the plates press together and become new metamorphic rocks.

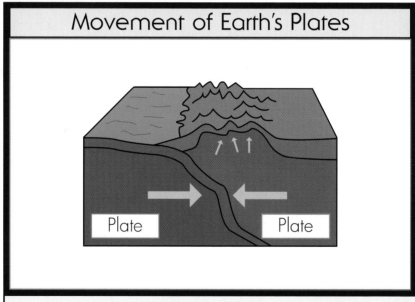

Movement of Earth's Plates

Plate   Plate

When two plates come together, the rocks within the plates are forced in different directions. Sometimes, this movement pushes the earth and rocks upward to form mountains.

Sometimes, we can see rock that has folded from movement in the earth's surface.

Making Different Rocks

Different amounts of heat and pressure make different rocks. Shale is a sedimentary rock. When it is heated and pressed, it becomes slate. With more heat and pressure, slate becomes schist. Slate and schist are metamorphic rocks.

The Fact Box

The more pressure and heat a rock has on it, the stronger it will become. With enough heat and pressure, the mineral graphite can turn into diamond! Diamond is the hardest matter found in nature.

Shale

Slate

Schist

Gneiss

With enough heat and pressure, shale can turn into gneiss.

Minerals in Rocks

Rocks are made of minerals. When rocks are pressed and heated, the minerals inside them change. Minerals in the rocks become more tightly packed. The minerals may also become larger.

A special light was used to take this picture of schist. The picture shows how the minerals in the rock formed in many different directions.

Additional heat and pressure turns schist into gneiss. New minerals with different shapes and sizes (seen here with a special light) were made when this gneiss was formed.

We can see the large minerals in some metamorphic rocks. Metamorphic rocks that were made with the most heat and pressure have large bands of light and dark minerals. Some minerals, like talc and graphite, are only found in metamorphic rocks.

The metamorphic mineral graphite is used as lead in pencils.

The metamorphic mineral talc is so soft, it is often used in baby powder.

Using Metamorphic Rocks

Metamorphic rocks are used for many different things. Marble is commonly used for buildings and sculptures. Slate is used to make tiles for roofs.

Some buildings have slate tiles on their roofs.

Marble was used to build the Taj Mahal in India.

Some of the oldest known rocks in the world are metamorphic rocks. They have been forming for about four billion years. Even today, under mountains and deep within Earth, metamorphic rocks are still forming.

The Fact Box

People have been using rocks for more than 2 million years. In Africa, scientists have found simple stone tools that are about 2.5 million years old.

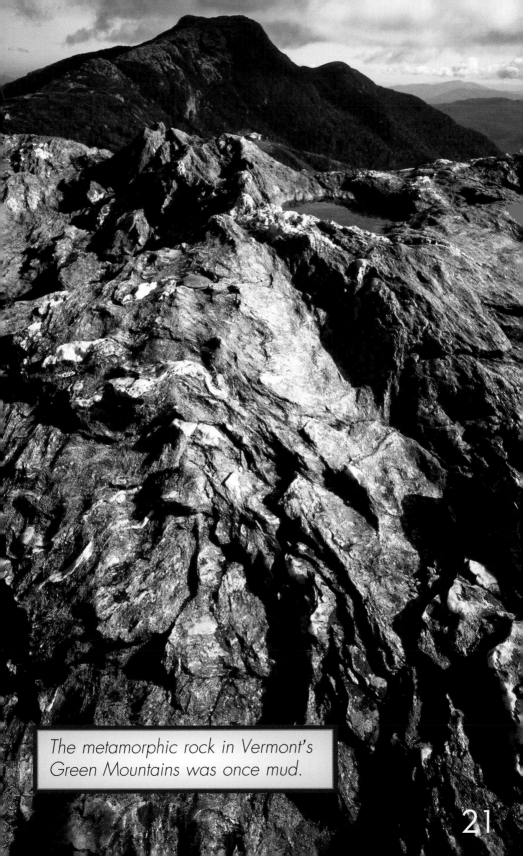

The metamorphic rock in Vermont's Green Mountains was once mud.

Glossary

igneous rocks (**ihg**-nee-uhs **rahks**) rocks made from magma or lava

lava (**lah**-vuh) hot, melted rock, or magma, that reaches Earth's surface

layers (**lay**-uhrz) one thickness or level of something that is on top of another

metamorphic rock (meht-uh-**mor**-fihk **rahk**) rock that is formed from other rocks

mineral (**mihn**-uhr-uhl) solid matter that comes from the earth

pressure (**prehsh**-uhr) the continued action of a weight or force

sculptures (**skuhlp**-chuhrz) figures that are shaped from stone, metal, wood, or other things

sediment (**sehd**-uh-muhnt) bits of rock, sand, dirt, and shells that settle in bodies of water, deserts, and valleys

sedimentary rocks (sehd-uh-**mehn**-tuhr-ee **rahks**) rocks that are formed by layers of sediment, which are being pressed together

surface (**ser**-fihs) the outside of something

Resources

Books
Metamorphic Rocks
by Melissa Stewart
Heinemann Library (2002)

Rocks and Minerals
by Tracy Staedter
Reader's Digest Children's Books (1999)

Web Sites
Due to the changing nature of Internet links, PowerKids
Press has developed an on-line list of Web sites related
to the subjects of this book. This site is updated regularly.
Please use this link to access the list:

http://www.powerkidslinks.com/ear/met/

Index

Word Count: 418

Note to Librarians, Teachers, and Parents

If reading is a challenge, Reading Power is a solution! Reading Power is perfect for readers who want high-interest subject matter at an accessible reading level. These fact-filled, photo-illustrated books are designed for readers who want straightforward vocabulary, engaging topics, and a manageable reading experience. With clear picture/text correspondence, leveled Reading Power books put the reader in charge. Now readers have the power to get the information they want and the skills they need in a user-friendly format.